JACK CARL KIEFER
COLLECTED PAPERS

Jack Carl Kiefer
Collected Papers
Supplementary Volume

Published with the co-operation of the
Institute of Mathematical Statistics
and edited by

Lawrence D. Brown
Ingram Olkin
Jerome Sacks
Henry P. Wynn

Springer-Verlag
New York Berlin Heidelberg Tokyo

Lawrence D. Brown
Department of Mathematics
Cornell University
Ithaca, NY 14853
U.S.A.

Ingram Olkin
Department of Statistics
Stanford University
Stanford, CA 94305
U.S.A.

Jerome Sacks
Department of Statistics
University of Illinois
Urbana, IL 61801
U.S.A.

Henry P. Wynn
The City University
Northhampton Square
London EC1V 0HB
England

AMS Subject Classifications: 01A75,62-03

Library of Congress Cataloging in Publication Data
(Revised for volume 4)
Kiefer, Jack, 1924-
Jack Carl Kiefer collected papers.
"Published with the co-operation of the Institute of
Mathematical Statistics."
Includes bibliographies.
Contents: 1. Statistical inference and probability,
1951–1963—2. Statistical inference and probability,
1964–1984—[etc.]—4. Supplementary volume.
1. Mathematical statistics—Collected works.
I. Brown, Lawrence D. II. Institute of Mathematical
Statistics. III. Title.
QA276.A12K54 1985 519.5 84-10598

Printed and bound by Quinn-Woodbine Inc., Woodbine, New Jersey.
Printed in the United States of America.

9 8 7 6 5 4 3 2 1

ISBN 0-387-96383-9 Springer-Verlag New York Berlin Heidelberg Tokyo
ISBN 3-540-96383-9 Springer-Verlag Berlin Heidelberg New York Tokyo

Contents

The commentaries in this volume provide reviews of selected papers from the three-volume Collected Papers of Jack Carl Kiefer (Springer-Verlag, 1985). The titles and page numbers of the papers listed below refer to these collected works.

Volume I

Volume II

COMMENTARY ON PAPERS [8] AND [20]

Zvi Galil[1]

Columbia University, Tel-Aviv University

In his masters' thesis from Massachusetts Institute of Technology in 1948, Kiefer proved the optimality of what is now called Fibonacci Search. It is a search for the maximum of a unimodal function on the unit interval. Allowing n function evaluations, one wants to find an interval, as small as possible, in which the maximum must lie. Several years later his advisor, Jack Wolfowitz, encouraged him to publish his result. This is paper [8]. This classic result is elegant and simple. We describe below a simplified version of Kiefer's proof.

The Fibonacci sequence is defined by $F_0 = F_1 = 1, F_{n+1} = F_n + F_{n-1}$ for $n > 1$. A function $f : [a, b] \to R$ is *unimodal* if there is an $x* \in [a, b]$ such that f is monotone increasing on $[a, x*]$ and monotone decreasing on $[x*, b]$. We pose the search problem in an equivalent form as follows: assume f is unimodal on $[0, L], L > 0$, and we are allowed to evaluate f at n points, $n \geq 2$.

Question: Can the value of $x*$ be pinned down to an interval of length ≤ 1?

Answer: Yes, if and only if $L < F_n$.

Proof: We use induction on n for both directions. In both cases the base ($n = 2$) is immediate.

The induction assumption for the 'if' part: assume $L < F_n$ and the value of f at $y = (F_{n-1}/F_n)L$ is known. Then using $n - 1$ evaluations of f we can pin down the value of $x*$ to an interval of length ≤ 1. To prove the induction step, assume it holds for $m = n - 1$ and evaluate f at $x = (F_{n-2}/F_n)L$. As a result we pin down the value of $x*$ to $[0, y], [x, L]$ or $[x, y]$. The last case (when $f(x) = f(y)$) cannot be worse than any one of the first two, so we ignore it. As a result we have an interval of length $L' = (F_{n-1}/F_n)L < F_{n-1}$, and we know the value of f at $y' = F_{n-2}/F_n = (F_{n-2}/F_{n-1})L'$. (In the second case we first shift the interval to the

[1] Work supported by NSF Grant MCS83-03139

origin.)

The induction assumption for the 'only if' part: If we can pin down the value of $x*$ to an interval of length ≤ 1 with at most m evaluations of f, then $L < F_m$. To prove the induction step, assume that it holds for all $m < n$. We first need to make the simple observation (used several times below and once above) that evaluating f at a point outside the interval or at an endpoint of the interval does not help. Assume the first two evaluations are at $x < y$.

Case 1: $x* \in [0, y]$. Since we find $x*$ in this new interval using $m - 1$ evaluations (including the one at x but excluding the one at y), we must have by induction $y < F_{m-1}$.

Case 2: $x* \in [y, L]$. Since we find $x*$ in this interval using $m - 2$ evaluations, $L - y < F_{m-2}$. Hence $L < F_{m-1} + F_{m-2} = F_m$.

It follows from the proof above that for every $\varepsilon > 0$ if we start with the unit interval and evaluate f at n points we can pin down the value of $x*$ to an interval of length $1/F_n + \varepsilon$, but not to an interval of length $1/F_n$.

Paper [20] considers several generalization of [8]. The search for the maximum of a unimodal function is of *second order*. Each decision is based on comparing two values of the function. The canonical example for first order search is the one where we look for a root of a monotone function. In this case the well known "binary search" is easily shown to be optimal.

Kiefer describes a third order search for an inflection point for an appropriate family of functions. He also points out the difficulty encountered when one tries to extend the Fibonacci search to higher dimensions. He also considers searches where evaluations are taken only at points of a lattice (in one and higher dimensions).

Kiefer also investigates the problem of estimating integrals of real functions. He gives optimal solutions for two classes of functions: monotone functions, and functions satisfying a Lipschitz condition with a given constant. (Here he minimizes the error of the computed integral.)

There have been many subsequent papers that considered similar search problems. One of the most successful attempts in treating the high dimensional search is due to Newman (1965) who found a notion analogous to unimodality. Newman's method uses $c_k \log n$ function evaluations to find the maximum on a lattice with $(n+1)^k$ points. This result is surprising and improves the obvious bound of $O(\log^k n)$ if Fibonacci search is used recursively. The latter bound has been observed and published by several authors whose treatment was marred by a lack of mathematical precision. Newman's method is optimal up to a constant factor in view of Kiefer's one- dimensional result.

Traub and Woźniakowski (1980) developed a general theory in which searches for maxima or for zeros, and approximations of integrals (and many other problems) are special cases. According to them, this theory had its inception in the work of Kiefer and of Sard and Nikolskij around 1950. The last two considered approximation of integrals in various settings. Traub and Woźniakowski (1980) contains an excellent annotated bibliography of the subject.

Reference

Newman, D. J. (1965). Location of the maximum on unimodal surfaces, *J. Assoc. Comp. Mach.*, 12, 395–398.

Traub, J. F. and Woźniakowski, H. (1980). *A General Theory of Optimal Algorithms*, Academic Press: New York.

COMMENTARY ON PAPER [4]

David Siegmund
Stanford University

Paper [4] is concerned with recursive estimation of the maximum (or minimum) of a function which can be determined experimentally up to some additive stochastic error. Kiefer, in his masters' thesis which later appeared as [8], had earlier considered the problem when there is no error. The method estimates the location of the maximum after $2n$ observations by a linear combination (depending on n) of the estimate after $2(n-1)$ observations and an estimate of the slope of the function at this previous estimate. The method was stimulated by that of Robbins and Munro (1951), who proposed a similar recursive scheme for estimation of the zero of a function observed with error.

Since 1952 more than a hundred papers and at least three books have been written on the subject of stochastic approximation. One notable contribution is Dvoretzky's theorem (1956), which showed that the results of both Kiefer-Wolfowitz and Robbins-Munro are special cases of an abstract stochastic approximation scheme.

Thus far, the most important applications of these ideas have been to on line identification and adaptive control of stochastic systems. Although recursive least squares estimators seem to converge more rapidly and hence to be more useful in practice, the simpler stochastic approximations methods are more easily analyzed and often provide important insights and tools for dealing with complex nonlinear control problems. Applications have been hindered by the lack of adequate stopping rules, a gap which has vexed all research in this area.

References

Dvoretzky, A. (1956). On stochastic approximation. *Proc. Third Berkeley Symp. Math. Statist. and Probability*, Vol. 1, 39–56, University of California Press.

Robbins, H. and Munro, S. (1951). A stochastic approximation method. *Ann. Math. Statist.*, 22, 400–407.

COMMENTARY ON PAPERS [9], [10], [17]

W. J. Hall
University of Rochester

In paper [9],[1] the authors point out some simple facts: Wald's sequential analyses may be done in continuous time. These facts contribute substantially to understanding the fundamental simplicity of Wald's sequential probability ratio test (SPRT) and the theory around it, and they are taken for granted today.

More specifically, all of the concepts, methods and results on sequential testing of two simple hypotheses, developed for sequences of independent and identically distributed random variables by Wald (1947), carry over to analogous stochastic processes in continuous time: an SPRT is simply described, with control of the error probabilities, formulae for the average stopping time, the optimality property, etc. Indeed, the authors point out that some aspects are simpler than in discrete time: some of the basic theorems of sequential analysis can, in continuous time, be proved as elementary consequences of basic facts in martingale theory; and in some important examples in continuous time problems of excess over stopping boundaries disappear or are simplified, compared with discrete time.

Conceptually most important is the case of testing hypotheses about the drift of a Wiener process (with known variance), one of the two examples treated in this paper. Because of the continuous-sample-path property, there is no "overshoot" of stopping boundaries, and various formulas—in particular, operating characteristic (equivalently, power) and average stopping time formulas—are available as exact expressions rather than as approximations, as derived by Wald for discrete time. The value of this today is largely two-fold: primarily, it provides a conceptually simple "canonical problem" of sequential-testing (see below)—one that is easy to understand and easy to derive results for, and which serves as an approximation to

[1] A correction to [9] appeared in *Ann. Math. Statist.*, 30 (1959), 1265.

many other problems. Secondly, it makes possible an alternative derivation of discrete-time approximations, thereby contributing further to their understanding and validity. Thirdly, it also provides a solution to a testing problem for Wiener processes.

All of the results above are pointed out (pages 256–8) without the necessity for writing down any formal mathematics.

The authors do two other things in this paper: (1) they demonstrate, by a brief argument (pages 258–9), that questions of testing hypotheses about the variance of a Wiener process do not arise—in contrast to the discrete-time case—since (in principle) the variance can be determined by observation of the process over any arbitrarily short time interval! (2) In their study of the Poisson process example (pages 259–64), they go beyond the Wald arguments regarding bounds on the OC and AST functions, and develop and solve differential-difference equations for them, thus providing exact formulas. (However, such formulas were obtained earlier by Burman (1946), by taking limits in formulas for binomial sampling in discrete time.) Tables based on these solutions appear in [17]; this problem of sequentially observing a Poisson process and testing hypotheses about its intensity was deemed (and is) of practical interest—hence, the tables.

Much of this material has been incorporated in the textbook of Ghosh (1970). More importantly, some of it plays an implicit role in much of the current writings on sequential testing. As already noted, testing about the drift of a Wiener process can be viewed as the canonical testing problem of sequential analysis—just as testing about the mean of a normal distribution is the canonical problem of non-sequential testing. Potential test statistics, for many statistical problems, are approximately normal, and so normal mean test concepts and procedures may be carried over. Similarly, many sequences of statistics are approximately (in the sense of weak convergence) Wiener processes, and hence test procedures for Wiener processes have broad applicability sequentially.

Ideas along these latter lines started with Bartlett (1946); later examples include Armitage (1957), Cox (1963), Hall & Loynes (1977), Lai (1976), Sen (1982), and Whitehead (1983). But this paper was the first to describe this "canonical test"—the SPRT for the drift of a Wiener process. (Variations on the SPRT for Wiener processes have become popular more recently;

see Anderson (1960), Hall (1980) and Whitehead (1983), for example.)

Another connecting link between the Wiener process SPRT of this paper and the discrete-time SPRT's of Wald is developed in Hall (1969)—by embedding the latter into the former, in Skorokhod fashion.

Decision-theoretic aspects, in particular the minimax character, of certain SPRT's in continuous time were later considered by Breakwell (1956) and by DeGroot (1960).

Paper [10] is concerned with decision-theoretic formulations of sequential estimation. Specifically, it presents four estimation problems about parameters of continuous-time stochastic processes for which it is shown that minimax estimations schemes are in fact non-sequential. Each example is treated by the so-called Bayes method: in brief, you find Bayes non- sequential procedures with constant posterior risks coinciding with their maximum risks. These examples are noted again in [19] where the invariance method of finding minimax solutions is developed.

References

Anderson, T. W. (1960). A modification of the sequential probability ratio test to reduce the sample size. *Ann. Math. Statist.*, 31, 165– 197.

Armitage, P. (1957). Restricted sequential procedures. *Biometrika*, 44, 9–26.

Bartlett, M. S. (1946). The large-sample theory of sequential test. *Proc. Camb. Phil Soc.*, 42, 239–244.

Breakwell, John V. (1956). Economically optimum acceptance tests. *J. Amer. Statist. Assoc.*, 51, 243–256.

Burman, J. P. (1946). Sequential sampling formulae for a binomial population. *Suppl. J. Roy. Statist. Soc.*, 8, 98–103.

Cox, D. R. (1963). Large sample sequential tests for composite hypotheses. *Sankhyā A*, 25, 5–12.

DeGroot, M. H. (1962). Minimax sequential test of some composite hypothese. *Ann. Math. Statist.*, 31, 1193–1200.

Ghosh, B. K. (1970). *Sequential Tests of Statistical Hypotheses.* Addison-Wesley: Reading, Mass.

Hall, W. J. (1969). Embedding submartingales in Wiener processes with drift, with applications to sequential analysis. *J. Appl. Prob.*, 6, 612–632.

Hall, W. J. (1980). Sequential minimum probability ratio tests. *Asymptotic Theory of Statistical Tests and Estimation.* (ed. M. Chakravarti), pp. 325–350. Academic Press: New York.

Hall, W. J. and Loynes, R. M. (1977). Weak convergence of processes related to likelihood ratios. *Ann. Statist.*, 5, 330–341.

Lai, T. L. (1978). Pitman efficiencies of sequential tests and uniform limit theorems in nonparametric statistics. *Ann. Statist.*, 6, 1027– 1047.

Sen, Pranab Kumar (1981). *Sequential Nonparametrics: Invariance Principles and Statistical Inference.* Wiley: New York.

Wald, Abraham (1947). *Sequential Analysis.* Wiley: New York.

Whitehead, J. (1983). *The Design and Analysis of Sequential Clinical Trials.* Ellis Horwood Ltd.: Chichester, UK.

COMMENTARY ON PAPERS [18] AND [37]

Gary Lorden

California Institute of Technology

The 1957 paper, [18], of Kiefer and Weiss and the follow-up by Weiss (1962) took up an important problem in sequential testing that became known as the Kiefer-Weiss problem. To test two separated hypotheses about a real parameter θ, it is natural to prescribe upper bounds on the error probabilities at two points, θ_{-1}, θ_1, and to seek to use a test that requires the least sampling among those satisfying the error probability constraints. The optimality property of the Sequential Probability Ratio Test (SPRT), (Wald and Wolfowitz (1948)) shows that a test of this type minimizes the expected sample size for both θ_{-1} and θ_1. It was widely appreciated, however, that the SPRT is much less efficient in minimizing the expected sample sizes for values of θ between θ_{-1} and θ_1, which are typically much larger. An earlier paper of Weiss (1953) defined generalized SPRT's, which allow the critical values for the likelihood ratio to depend on the number of observations taken. Part of [18] is devoted to admissibility of these GSPRT's and to a complete class result for them. An earlier result by Sobel (1952) concerned a more general testing problem for the case of exponential families. Current knowledge of admissibility for sequential tests covers broader types of problems (see the references to Berk, Brown, Cohen, Samuel-Cahn, and Strawderman), though not yet the Student's t problem. These results have been obtained by similar methods, relying on characterizations of Bayes solutions.

The most important part of [18] concerns the structure of those tests satisfying error constraints at θ_{-1} and θ_1 that solve the following:

(i) (Modified Kiefer-Weiss Problem) To minimize the expected sample size at a (given) third point θ_0.

(ii) (Kiefer-Weiss Problem) To minimize the maximum expected sample size (over all θ).

Kiefer and Weiss showed that, under a reasonable set of assumptions, the solutions of the first problem are GSPRT's whose critical values converge monotonically to a truncation point. The second problem is discussed in a remark at the end of the paper, where it is noted that a test solving (i) also solves (ii) if its maximum expected sample size is at $\theta = \theta_0$. In symmetric cases, such as testing with equal error constraints the mean of a normal distribution and testing $p = p_0$ versus $p = 1 - p_0$ for the binomial parameter, this observation leads to a reduction of problem (ii) to problem (i) with θ_0 midway between θ_{-1} and θ_1. These symmetric cases were studied in the paper of Weiss (1962), where the truncation points are determined and the possibility (and difficulty) of computing optimal tests by "backward induction" from the truncation point (as in (Wald, 1950)) is discussed.

A printer's error caused the interchanging of most of pages 70–72 of [18] with most of pages 14–17 of the same issue of the *Annals*. Readers should also note that, although Assumptions A, B, and C of Section 4 are intended to provide a domain of applicability broader than the usual one- dimensional exponential families, they achieve little. As pointed out in Remark 7 of Lorden (1980), under mild regularity conditions it is only exponential families for which modified Kiefer-Weiss solutions are GSPRT's. The two examples in Lemma 4.1 of [18] have in common a very special structure in which the likelihood ratios take values in $(0,1) \bigcup \{+\infty\}$.

Both Kiefer-Weiss problems remain unsolved, although much more has been learned in recent years about approximate and exact solutions. Lai (1973) obtained information about exact optimal boundaries in the Wiener process case. Lorden (1980) obtained additional information about the structure of modified Kiefer-Weiss solutions in general, including the determination of truncation points and asymptotes to the stopping boundaries. The most important results, however, concern approximate solutions. It has been known since Anderson's (1960) important paper that, at least in the symmetric normal case, simple "triangular boundaries" come close to attaining the minimax expected sample size. Lorden (1976) showed in a general (not necessarily symmetrical) setting that the modified problem is solved within $o(1)$ as the error probabilities go to zero by a natural type of test, the 2-SPRT, which simultaneously carries out one- sided SPRT's of θ_0 vs. θ_{-1} and θ_1; numerical computations for the symmetric normal case showed that 2-SPRT's come within 1% of the optimum at typical significance levels. Huffman (1983) gives a recipe in the case of exponential families for choosing

2-SPRT's (i.e. choosing θ_0 and the critical values) that attain the minimax expected sample size asymptotically to second order. To investigate the small-sample efficiency of his method, Huffman carried out the program suggested in (Lorden, 1980) to find Kiefer-Weiss solutions as a by-product of the calculation of modified Kiefer-Weiss solutions, which nowadays is not so formidable a computational task. He found in the case of the negative exponential distribution that small-sample efficiencies of about 98% are typically attained by his prescription. Recently, Lai (1981) has obtained impressive asymptotic results for both invariant SPRT's and invariant 2-SPRT's, showing in the case of the latter that the appropriate generalization of the modified Kiefer-Weiss problem is solved asymptotically as the error probabilities go to zero.

Paper [37] made rich and important contributions to the theory of sequential analysis and the the subject of sequential design initiated by Chernoff (1959). Chernoff's work established the fundamental ideas of asymptotic optimality theory for sequential procedures using small cost c per observation; he also allowed for the "design" feature of choosing among available experiments for each successive observation. Chernoff's formal results were limited to problems with finitely many states of nature. Bessler (1960) and Albert (1961) extended the developments to infinite cases. In Schwarz's striking paper (1962), the notion of asymptotic optimality was applied to the testing of separated hypotheses about the parameter θ of a one-dimensional exponential family, leading to his famous "asymptotic shapes", which are the continuation regions of tests based on generalized likelihood ratio statistics. The very readable and informative introduction in [37] discusses the relationship of their investigation to these previous results, as well as those of [19] and Anderson (1960); in addition, it gives a wonderfully enlightening discussion of the asymptotic performance of a variety of proposed tests for the normal mean. The goals of Kiefer and Sacks' investigation are clearly stated as

(i) generalizing Schwarz's results.

and

(ii) extending and simplifying the sequential design results of Chernoff, Bessler and Albert.

These will be discussed separately.

The sequential testing problems are formulated quite generally as sequential decision problems for two or more separated hypotheses, allowing for indifference and semi-indifference regions. Assumptions are made about compactness of the parameter space and continuity properties involving the first and second moments of the log-likelihood ratios. These assumptions permit the problem to be treated essentially as though the parameter space is finite, although considerable virtuosity is required (including generalizations of fluctuation theory results of Erdös (1949) and Spitzer (1956)). The main results state that certain tests are asymptotically Bayes as the cost c per observation goes to zero, the tests being essentially like the ones Schwarz studied, which stop when the a posteriori risk is less than c. It is shown that the tests based on a particular prior are asymptotically Bayes with respect to any prior having the same support. Still, the tests require numerical evaluation of mixtures of likelihoods at each stage, unlike Schwarz's "shapes", which require only evaluation of generalized likelihood ratio statistics (particularly easy in Schwarz's exponential family setting). The mathematical advantage of the mixture approach is that it leads by a simple standard argument (Lemma 1) to nice upper bounds on the risk due to error. Good uniform upper bounds on the error probabilities of sequential generalized likelihood ratio tests are still not known in a general setting.

The sequential design results of Kiefer and Sacks are based upon their non-design theory, together with a device based upon an idea of Wald (1951) in a simpler problem of estimation: take a "preliminary sample" (whose size goes to infinity, but more slowly than the overall sample size), use it to estimate the true state of nature, then continue according to a pattern of choices of experiments that would be approximately optimal if the estimated state of nature were true. Such a design pattern can be chosen so that observations are taken in blocks from a finite number of experiments, thus requiring a minimum of calculations and switching of experiments. Previous recipes for general problems had called for randomized choice of experiments, with randomization probabilities varying from stage to stage as the estimated true state of nature changes.

A good deal more is now known about asymptotic optimality of sequential hypothesis tests. Using stronger separation assumptions Lorden (1967) strengthened the results of [37] from asymptotic optimality within $O(c \log c)$ to $o(c)$ (the cost of a bounded number of observations). Lorden's result was based on showing that Bayes procedures continue sampling when the stopping risk exceeds a multiple M of the cost c. A similar approach was used by

Pollak (1978) to obtain a minimax type of result for open-ended tests of exponential families. In Schwarz's context and in the case of finitely many states of nature, Lorden (1977, 1984) improved the optimality result to $o(c)$ using generalized likelihood ratio procedures. Recently, R. Lerche (to appear) has successfully applied similar method to open-ended testing of *unseparated* hypotheses in exponential families using a certain cost structure (essentially, constant cost per unit of information).

In the area of sequential design, less progress has been made along the lines of the Kiefer and Sacks investigation. Asymptotic theory of this type has not shed much light on important practical problems such as the comparison of treatments in clinical trials. The best recipes known are derived by more direct analysis of specific problems, and even operating characteristics and comparisons of procedures are hard to compute. Some recent results of Keener (1984) and Lalley and Lorden (to appear) offer improved asymptotic optimality in the case of finitely many states of nature, but generalization of these results appears to be difficult.

References

Albert, A. E. (1961). The sequential design of experiments for infinitely many states of nature. *Ann. Math. Statist.*, 32, 774–799.

Anderson, T. W. (1960). A modification of the sequential probability ratio test to reduce the sample size. *Ann. Math. Statist.*, 31, 165– 197.

Berk, R. H., Brown, L. D. and Cohen, A. (1981). Bounded stopping times for a class of sequential Bayes tests. *Ann. Statist.*, 9, 834–845.

Berk, R. H., Brown, L. D., and Cohen, A. (1981). Properties of Bayes sequential tests. *Ann. Statist.*, 9, 678–682.

Bessler, S. A. (1960). Theory and applications of the sequential design of experiments, k-actions, and infinitely many experiments. Technical Report No. 55, Department of Statistics, Stanford University.

Brown, L.D., Cohen, A., and Samuel-Cahn, E. (1983). A sharp condition for admissibility of sequential tests—necessary and sufficient conditions for admissibility of SPRT's. *Ann.*

Statist., 11, 640–653.

Brown, L. D., Cohen, A. and Strawderman, W. E. (1979). Monotonicity of Bayes sequential tests. *Ann. Statist.*, 7, 1222–1230.

Brown, L. D., Cohen, A. and Strawderman, W. E. (1980). Complete classes for sequential tests of hypotheses. *Ann. Statist.*, 8, 377– 398.

Brown, L. D. and Cohen, A. (1981). Inadmissibility of large classes of sequential tests. *Ann. Statist.*, 9, 1239–1247.

Cohen, A. and Samuel-Cahn, E. (1982). Necessary and sufficient conditions for bounded stopping time of sequential Bayes tests in one- parameter exponential families. *Sequential Analysis, Comm. in Statistics*, 1, 89–99.

Chernoff, H. (1959). Sequential design of experiments. *Ann. Math. Statist.*, 30, 755–770.

Erdös, P. (1949). On a theorem of Hsu and Robbins, *Ann Math. Statist.*, 20, 286–291.

Huffman, M. D. (1983). An efficient approximate solution to the Kiefer- Weiss problem. *Ann. Statist.*, 11, 306–316.

Keener, R. W. (1984). Second order efficiency in the sequential design of experiments. *Ann. Statist.*, 12, 510–532.

Lai, T. L. (1973). Optimal stopping and sequential tests which minimize the maximum expected sample size. *Ann. Statist.*, 1, 659– 673.

Lai, T. L. (1981). Asymptotic optimality of invariant sequential probability ratio tests. *Ann. Statist*, 9, 318–333.

Lorden, G. (1967). Integrated risk of asymptotically Bayes sequential tests. *Ann. Math. Statist.*, 38, 1399–1422.

Lorden, G. (1976). 2-SPRT's and the modified Kiefer-Weiss problem of minimizing and expected sample size. *Ann. Statist.*, 4, 281–291.

Lorden, G. (1977). Nearly-optimal sequential tests for finitely many parameter values. *Ann. Statist.*, 5, 1–21.

Lorden, G. (1980). Structure of sequential tests minimizing an expected sample size. *Z. Wahrsch. Verw. Gebiete*, 51, 291–302.

Lorden, G. (1984). Nearly optimal sequential tests for exponential families. (to appear in *Ann. Statist.*).

Pollak, M. (1978). Optimality and almost optimality of mixture stopping rules. *Ann. Statist.*, 6, 910–916.

Schwarz, G. (1962). Asymptotic shapes of Bayes sequential testing regions. *Ann. Math. Statist.*, 33, 224–236.

Sobel, M. (1952). An essentially complete class of decision functions for certain standard sequential problems. *Ann. Math. Statist.*, 23, 319–337.

Spitzer, F. (1956). A combinatorial lemma and its application to probability theory. *Trans. Amer. Math. Soc.*, 82, 332–339.

Wald, A. and Wolfowitz, J. (1948). Optimum character of the sequential probability ratio test. *Ann. Math. Statist.*, 19, 326–339.

Wald, A. (1950). *Statistical Decision Functions.* Wiley: New York.

Wald, A. (1951). Asymptotic minimax solutions of sequential point estimation problems. *Proc. Second Berkeley Symp. Math. Statist. and Probibility*, 1–11, University of California Press.

Weiss, L. (1953). Testing one simple hypothesis against another. *Ann. Math. Statist.*, 24, 273–281.

Weiss, L. (1962). On sequential tests which minimize the maximum expected sample size. *J. Amer. Statist. Assoc.*, 57, 551–566.

COMMENTARY ON PAPER [16]

Michael Woodroofe
Rutgers University and University of Michigan

This clasic paper makes several important contributions. The most obvious is to establish the consistency of the maximum likelihood estimator of a structural parameter θ when there are i.i.d. incidental parameters $\alpha_1, \alpha_2, \cdots$ with a common distribution function G (in the notation of the paper). In the process, the paper establishes the consistency of the non-parametric, maximum likelihood estimator of G. Finally, buried deep in the paper, there are insightful remarks which anticipate and stimulate future developments.

The first, most obvious contribution, has several implications. Thus Example 2, regression with both variables subject to error, shows that maximum likelihood estimation works in this problem, if the independent variables are i.i.d. Lehmann (1983, pp. 450–51) describes this example and refers to Kiefer and Wolfowitz' work on it. Example 1, estimating a structural location parameter in the presence of incidental scale parameters is especially interesting. In principle, it provides a method for constructing adaptive estimators of a location parameter, when the possible distributions of X are scale mixtures of normal distributions (with $G(\varepsilon) = 0$ for some $\varepsilon > 0$): estimate θ and G jointly by maximum likelihood; then $\hat{\theta}$ is the maximum likelihood estimator in the model with $G = \hat{G}$.

Even when θ is absent from the model, the problem of estimating G is an important one. There is current interest in this problem, since the model arises in empirical Bayesian problems and compound decision problems. The example of Neyman and Scott (1958) shows that estimating the incidental parameters by maximum likelihood may lead to problems. If G were known and treated as a prior distribution, then one could estimate the incidental parameters by Bayesian methods; and Kiefer and Wolfowitz' result shows that the maximum likelihood estimator of G is consistent, provided that G is identifiable and the other regularity conditions are satisfied. There are serious difficulties in implementing these potential applications, since the maximum likelihood estimator of G is difficult to compute. See Lindsay (1983) for recent work on this problem.

A brief remark (on page 893) on the proper definition of the maximum likelihood estimator in the undominated case anticipated substantial subsequent interest in non-parametric maximum likelihood estimation. According to this definition, the empirical distribution function is the maximum likelihood estimator of the distribution function of a simple random sample, and the product-limit estimator is the non-parametric maximum likelihood estimator of the distribution of survival times when censored survival times are observed. See Miller (1981, pp. 57–59) for the latter derivation. The remark is important, since it provides a method for defining descriptive statistics (maximum likelihood estimators of a distribution) when the data are not a simple random sample.

Kiefer and Wolfowitz draw heavily on previous work of Wald for both motivation and technique. Wald (1940) had constructed consistent estimators of regression parameters when both variables were observed with error; Wald (1948) considered maximum likelihood estimation of an increasing number of parameters; and Wald (1949) developed a general method for proving the consistency of maximum likelihood estimators of a finite number of parameters. Kiefer and Wolfowitz used this method effectively in their paper. They also contributed to it by noting that Wald's main integrability condition may be satisfied when observations are taken in pairs, even if it is not satisfied when observations are taken singly. This remark was used by Perlman (1972) in obtaining necessary and sufficient conditions for consistency.

The paper has had a substantial impact. The *Science Citation Index* lists over fifty references to it during the years 1961–1983.

References

Lehmann, E. (1983). *Theory of Point Estimation.* Wiley: New York.

Lindsay, B. (1983). The geometry of mixtures of likelihoods: general theory. *Ann. Statist.*, 11, 86–94.

Miller, R. (1981). *Survival Analysis.* Wiley: New York.

Neyman, J. and Scott, E. L. (1948). Consistent estimates based on partially consistent observations. *Econometrica*, 16, 1–32.

Perlman, M. (1972). On the strong consistency of approximate maximum likelihood estimators. *Proc. Sixth Berkeley Symp. Math. Statist. and Probability*, 263–281.

Wald, A. (1940). The fitting of straight lines if both variables are subject to error. *Ann. Math. Statist.*, 11, 284–300.

Wald, A. (1948). Estimation of a parameter when the number of unknown parameters increases indefinitely with the number of observations. *Ann. Math. Statist.*, 19, 220–227.

Wald, A. (1949). Note on the consistency of the maximum likelihood estimate. *Ann. Math. Statist.*, 20, 595–601.

COMMENTARY ON PAPER [19]

Lawrence D. Brown[1]
Cornell University

This paper presents a comprehensive Hunt-Stein theorem. Such a theory guarantees the existence under suitable regularity conditions of an invariant minimax procedure. In many circumstances it is relatively easy to determine a procedure which is minimax within the class of invariant procedures. A Hunt-Stein theorem—when it applies—shows that this procedure is also minimax within the class of all procedures.

History: The Hunt-Stein theorem has a notable history of non- publication. This history up to 1957 is fully described in Section 1 of [19]. The first such (correct) result was due to Hunt and Stein and was never published. Important generalizations of the Hunt-Stein theorem appeared in Peisakoff's thesis (1950), but were never published elsewhere. Kiefer's paper, in any case, greatly improved Peisakoff's results; and it still stands as the most comprehensive treatment of the Hunt Stein theorem in the statistical literature.

There is now a method of proof for a general Hunt-Stein theorem which is more elegant and direct than the proof Kiefer uses. This method is apparently due to Huber but there is no complete published account of it. For the special problem of hypothesis testing Huber's argument is sketched in [45], and presented more fully in Banyaklef (1971) and in Bondar and Milnes (1981). Huber's method will be sketched later in this review.

Wesler (1959), Portnoy (1972), and Zehnwirth (1977) also treat the general Hunt-Stein theorem. Their results can all be derived as special cases of Kiefer's. Lehmann (1959) contains a very readable proof of a Hunt-Stein theorem for hypothesis testing.

Assumptions: The main theorem is stated and proved in Section 3. The assumptions for this theorem are collected in Section 2 as Assumptions 1–5 plus a measurability assumption

[1] Work supported in part by N.S.F. grant MCS 3200031

contained in the second paragraph of the section. These assumptions are stated in maximum generality and then refined and explained by subsequent statements of more easily verified conditions which imply these assumptions. Developments following Kiefer's paper enable a clearer interpretation of these assumptions, which I will now try to sketch.

The assumptions may be separated into four groups: the measurability assumptions, Assumptions 1 and 2, Assumptions 3 and 4, Assumption 5. The first three groups may be thought of as primarily technical. Mainly they eliminate the trivial and the pathological exceptions to the theorem. Assumption 5 is the basic assumption and so will be discussed first.

Assumption 5: Assumptions on the group, G, of this character appear in all current general Hunt-Stein theorems. As noted in Kiefer's Counterexample D the theorem fails without such an assumption. The assumption is not satisfied if for example the group is the multiplication group of all non-singular $n \times n$ matrices ($n \geq 2$). This group appears in MANOVA and in similar multivariate problems. There are statistical problems invariant under this group in which no minimax invariant estimator exists. Examples are given in Lehmann (1959) and in Stein (1955). On the other hand there are problems invariant under such a group in which a minimax invariant procedure does exist. See, for example, [36]. As of now there is no general result available for such situations; each problem must be treated individually, and most are still unsolved. (It is even shown in Portnoy (1972) that for any reasonable group not satisfying Assumption 5 there exists a problem invariant under this group for which no invariant minimax procedure exists. See also Bondar and Milnes (1981).)

There is a simple way of characterizing all topological groups which satisfy Assumption 5. The assumption is satisfied if and only if the group is solvable to a compact factor, i.e., there is a sequence $G = G^{(0)}, G^{(1)}, \ldots, G^{(2n)}$ with $G^{(2k+1)}$ a normal subgroup of $G^{(2k)}$, $G^{2k+2} = G^{(2k)}/G^{(2k+1)}$, $k = 0, \ldots, n-1$, and with $G^{(2n)}$ compact. Note that if G^{2n-2} is commutative, so that G is solvable, one takes $G^{2n-1} = G^{2n-2}$ and gets $G^{(2n)} = \{e\}$ which is compact. The sufficiency of this solvability condition is nearly proved by applying the arguments in Kiefer's Condition 5d, plus the remarks in his Conditions 5a and 5b. Kiefer realized this in the early 1960's after hearing of Huber's proof in which this solvability condition is the pivotal assumption. The equivalence of a variety of conditions including Assumption 5 and the above solvability condition is proved in Bondar and Milnes (1981).

The Measurability Assumption: This assumption includes not only standard measurability conditions, but also the assumption of existence of a (left) invariant measure μ. For most possible applications this assumption is innocuous. However in various nonparametric situations one encounters the group of continuous strictly monotone transformations of the line. This is not a locally compact topological group and does not have an invariant measure. (It also does not seem to satisfy any appropriate version of Assumption 5 such as the solvability condition mentioned above.) Thus such nonparametric problems are outside the scope of Kiefer's paper and to date very little is known about the existence of invariant minimax procedures for such problems.

The measurability assumptions here are similar to those required to prove that to every almost invariant procedure there is an equivalent invariant procedure. See Lehmann (1959, p. 225), Berk and Bickel (1968), and Berk (1970). Such a step is explicitly required by other proofs of Hunt-Stein theorems such as those of Lehmann and Huber mentioned above.

Assumptions 1 and 2: In most applications Condition 2b′ is satisfied. As noted, this implies Assumptions 1 and 2. (In brief, Condition 2b′ requires that $\mathcal{Y} = G/M \times \mathcal{Z}, g(y, z) = (gy, z)$, and $md = d$ for all $m \varepsilon M, d \varepsilon D$. This condition is extensively discussed in Wijsman (1967) and Bondar (1975).) Condition 2b is a generalization of 2b′ which is useful in a few problems. It is hard to see how Assumptions 1 and 2 can be satisfied in any realistic settings unless Condition 2b is satisfied.

Assumptions 1 and 2 have two interesting simple consequences. They guarantee that an invariant procedure exists. They also require that the sample space be at least as large as the group of transformations. This eliminates the situation of Example C (p. 587)—the, "don't you name the real number I name" game—in which $\mathcal{X} = R$ but the group G consists of all affine transformations, and thus can be represented as $R \times (0, \infty)$.

Assumptions 3 and 4: Assumption 3 is that the weak minimax theorem holds for the set $\mathcal{D}_I$ of invariant procedures. Assumption 4 is a related continuity condition, and is hard to verify directly. Condition 4b which implies Assumption 4 is still somewhat awkward. No special conditions are given which imply Assumption 3. In fact the general decision theoretic approach of LeCam (1955) yields accessible conditions which imply both these assumptions when $\mathcal{D}$ is the class of all nonrandomized procedures, as follows:

Assumption LC: Assume $\mathcal{F}$ is a dominated family of distributions, $W(F,(y,z),d) = W(F,d)$, (This may be generalized using Brown (1977).), D is a locally compact second countable space, $W(F,\cdot)$ is lower semicontinuous,

$$\sup\{\inf\{W(F,d) : d \notin C\} : C \subset D, C \text{ compact}\} = \sup\{W(F,d) : d \varepsilon D\} \;\forall\; F \varepsilon \mathcal{F}, \tag{1}$$

and that there exists an invariant procedure. Assume also that the transformations $g : D \to D$ are continuous.

Under the assumptions let $D^* = D \bigcup \{\infty\}$ be the one point compactification of D. Define the action of G on D^* in the obvious manner, in which $g\infty = \infty \;\forall\; g \varepsilon G$. Let $W(F,\infty)$ be the value defined in (1). Let $\mathcal{D}^*$ denote the set of all (randomized) decision functions on D^*. Then, according to LeCam, $\mathcal{D}^*$ is compact and the maps $\delta \to r^b_\delta(\xi), b \leqq \infty$, are lower semi-continuous whenever ξ is discrete (and for more general ξ by Brown (1977)). It is not hard to verify that $\mathcal{D}^*_I$ is closed, and that to each $\delta^* \varepsilon \mathcal{D}^*_I$ there corresponds a $\delta \varepsilon \mathcal{D}_I$ with

$$r^b_\delta(\xi) \leq r^b_{\delta^*}(\xi) \;\forall \xi. \tag{2}$$

Kiefer's Assumptions 3 and 4 then follow via familiar arguments.

Note that the above yields the existence of minimax procedures relative to $\mathcal{D}$ or to $\mathcal{D}_I$. This is presumably what Kiefer was alluding to in his remarks in the middle paragraph on p. 589.

Huber's proof of the Hunt-Stein theorem: Huber's proof requires Assumption LC, above, plus an appropriate version of Kiefer's Assumption 5. It does not explicitly require an analog of Kiefer's Assumptions 1 and 2.

One defines the group action on $\mathcal{D}$ by $g\delta(x,\Delta) = \delta(gx, g\Delta)$. It can be verified that G acts continuously on $\mathcal{D}^*$ and that the set $\mathcal{M}^* \subset \mathcal{D}^*$, say, of minimax procedures is closed in $\mathcal{D}^*$ and is an invariant set under the action of G. Consequently there is a fixed point $\delta^* \varepsilon \mathcal{M}^*$. (See Bondar and Milnes (1981).)

To say that δ^* is a fixed point means, equivalently, that δ^* is almost invariant. So, δ^* has an equivalent invariant version. (See Berk and Bickel (1968) and Berk (1970).) Let $\delta \varepsilon \mathcal{D}_I$ be the procedure described by (2), above. Then δ is the desired invariant minimax procedure.

Note where Huber's method eliminates the paradox of the "don't you name the real number I name" game. It is eliminated by the lower semi continuity assumption on $W(F, \cdot)$. Kiefer notes on p. 586–587 that such an assumption does indeed suffice to eliminate the paradox; however, in his approach, it is instead eliminated by Assumption 1 and 2.

Sequential problems: An important feature and innovation of this paper is the treatment of sequential problems of a very general nature. The general sequential formulation on p. 589–592 by itself is of interest. It allows at each stage a choice of whether to terminate sampling and make a terminal decision or whether to continue sampling, and, if so, which variable(s) to observe at the next stage. Many elements of this formulation first appeared in Kiefer's thesis [104] and in [2,3]. A similar formulation appear in LeCam (1955) and a generalization appears in Brown (1977).

The general sequential Hunt-Stein Theorem: It should be emphasized that the Hunt-Stein theorem of Section 3 usually applies directly to the general sequential problem. Kiefer notes this in remarks mid p. 591 and in (*xv*) on p. 597, but many readers have overlooked these remarks.

The Hunt-Stein theorem yields the existence of a minimax invariant rule among those taking enough observations to guarantee the existence of an invariant rule. (See p. 591 and Example B p. 587.) Thus both the stopping (or continuation) rule and the terminal decision rule must be invariant. This is a significant simplification when searching for a minimax rule. However it still usually leaves the determination of the minimax stopping rule as a very difficult problem. This fact is noted in remark (*xv*) on p. 597. This explains Kiefer's interest in establishing a specialized condition under which the minimax invariant stopping rule can be easily calculated.

The specialized sequential theorem: The conclusion of this theorem (p. 593) is that an invariant minimax procedure exists which is actually a fixed sample size procedure. When such a procedure is known to exist it is then relatively easy to calculate. Special cases of this result were previously described in [5] and [10]. Further examples are given on p. 596–597.

In addition to the assumptions for the fixed sample size case the key assumption here is Assumption 6. (The theorem hypothesizes also Assumption NR (p. 579). It is not clear whether this assumption can be relaxed.) Assumption 6 requires that there be a fixed dimensional sufficient statistic, T_i, for each stage, i, of the process and that, at each stage, the conditional distribution of the maximal invariant given T_i should be independent of the value of T_i. This last condition is the pivotal assumption which enables the special conclusion of the theorem.

It is also assumed (p. 589) that the observations are independent. This assumption is not required if the T_i are assumed to also be transitive. (See Bahadur (1954) and Hall, Wijsman, and Ghosh (1965).)

Restricted parameter spaces: The last paragraph of Section 3 observes that the basic theorem is often valid when $\mathcal{F}$ is only a suitably large subset of $\{gF_0 : g\varepsilon G\}$. This suggests the possibility of a general theorem of this nature. One possibility would be to define and consider problems invariant under a semigroup G'. The proof of Section 3 and the examples on p. 589 suggest the validity of a general result of this nature. (It is also suggested by the fact that the fixed point theorem used in Huber's approach is valid also for semigroups.)

Continuous time: The final section of the paper treats sequential problems with continuous time. Such problems were first considered in [9, 10]. What are given here are minimax results for a few special, but important, problems in continuous time. A general reuslt is also briefly sketched. Note also part II of this section which considers one special problem in which the group acts on the time axis rather than the sample space.

References

Bahadur, R. (1954). Sufficiency and statistical decision functions. *Ann. Math. Statist.*, 25, 423–462.

Benyaklef, M. (1971). Statistical invariance under groups and optimum procedures. University of California, Berkeley. Doctoral dissertation.

Berk, R. H. (1967). A special group structure and equivariant estimation. *Ann. Math. Statist.*, 38, 1436–1445.

Berk, R. and Bickel, P. (1968). On invariance and almost invariance. *Ann. Math. Statist.*, 39, 1573–1576.

Berk, R. H. (1970). A remark on almost invariance. *Ann. Math. Statist.*, 41, 733–735.

Bondar, J. V. (1975). Borel cross sections and maximal invariants. *Ann. Statist.*, 4, 866–877.

Bondar, J. V. and Milnes, P. (1981). Amenability: A survey for statistical applications of Hunt-Stein and related conditions on groups. *Z. Wahrsch. Verw. Gebiete.*, 57, 103–128.

Brown, L. (1977). Closure theorems for sequential design processes. *Statistical Decision Theory and Related Topics, II*, (ed. by S. S. Gupta and D. S. Moore), 57–92. Academic Press: New York.

Hall, W. J., Wijsman, R. A., and Ghosh, J. K. (1965). The relationship between sufficiency and invariance with applications in sequential analysis. *Ann. Math. Statist.*, 36, 575–614.

LeCam, L. (1955). An extension of Wald's theory of statistical decision functions. *Ann. Math. Statist.*, 26, 69–81.

Lehmann, E. L. (1959). *Testing Statistical Hypotheses*, John Wiley and Sons: New York.

Peisakoff, M. P. (1950). Transformation parameters. Princeton University. Doctoral dissertation.

Portnoy, S. (1972). On fundamental theorems in statistical decision theory. Lecture notes, Harvard University.

Stein, C. (1955). On tests of certain hypotheses invariant under the full linear group (abstract). *Ann. Math. Statist.*, 26, 769.

Wesler, O. (1959). Invariance theory and a modified minimax principle. *Ann. Math. Statist.*, 30, 1–20.

Wijsman, R. A. (1967). Cross sections of orbits and their application to densities of maximal invariants. *Proc. Fifth Berkeley Symp.. Math. Statist. and Probability*, 1, 389–400.

Zehnwirth, B. (1977). W^*–compactness of the class of substatistical decision rules with applications to the generalized Hunt- Stein theorem. Macquarie University School of Economic Studies Technical Report No. 135.

COMMENTARY ON PAPERS [12], [14]

N. U. Prabhu
Cornell University

In these papers the authors develop the theory of many-server queues with arbitrary distributions of inter-arrival times and service times, and the queue discipline, "first come, first served". Such a system is denoted as GI/G/s in the notation of D. G. Kendall (1951, 1954). Lindley (1952) and Smith (1953) further advanced this theory by investigating waiting times in single server queues (the case $s = 1$). The authors extend Lindley's results to the general case $s \geq 1$. To put their results in a proper perspective, suppose that successive customers $C_0, C_1, C_2, \ldots$ arrive at the epochs $t_0(= 0), t_1, t_2, \ldots$ and demand service times $v_1, v_2, v_3, \ldots$. The inter-arrival times are then $u_k = t_k - t_{k-1} (k \geq 1)$. It is assumed that $\{u_k\}$ and $\{v_k\}$ are two independent sequences of independent and identically distributed random variables with distribution functions (d.f.) A and B respectively. Denoting $E(u_k) = a$ and $E(v_k) = b$ we assume that $0 < a < \infty, 0 < b < \infty$. There are s servers $M_1, M_2, \ldots, M_s$. Each arriving customer goes to one of the free servers (for definiteness the one with the smallest index) if there is one such, or otherwise waits in the queue. The traffic intensity of the system is defined as $\rho = b/as (0 < \rho < \infty)$. Let W_n be customer C_n's waiting time; that is, W_n is the time the customer arriving at the epoch T_n will have to wait for the commencement of his service. To obtain an expression for W_n let us denote by d_{nj} the time at which server M_j completes his service on the last of the customers $C_0, C_1, \ldots, C_{n-1}$; also, let $w_{nj} = d_{nj} - t_n (1 \leq j \leq s)$. If $w_{nj} > 0$ then w_{nj} is server M_j's remaining workload at time t_n, while if $w_{nj} \leq 0$ then $|w_{nj}|$ is server M_j's idle time in progress at t_n. For any real number x let us denote $x^+ = \max(0, x)$. Then $w_{nj}^+ (1 \leq j \leq s)$ represent the earliest times the s-servers will be available to customer C_n. Actually his waiting time is given by

(1) $$W_n = \min\{w_{n1}^+, w_{n2}^+, \ldots, w_{ns}^+\}.$$

Let us go one step further and define $(W_{n1}, W_{n2}, \ldots, W_{ns})$ to be the quantities on the right side of (1) arranged in ascending order; thus C_n's waiting time is $W_n = W_{n1}$. His service begins at the epoch $t_n + W_{n1}$ and is completed at $t_n + W_{n1} + v_{n+1}$. The next arrival C_{n+1}'s waiting time W_{n+1} can therefore be expressed as

$$W_{n+1} = \max\{0, \min(t_n + W_{n1} + v_{n+1},\ t_n + W_{n2}, \ldots, t_n + W_{ns}) - t_{n+1}\},$$

or

$$W_{n+1} = \min\{(W_{n1} + v_{n+1} - u_{n+1})^+,\ (W_{n2} - u_{n+1})^+, \ldots, (W_{ns} - u_{n+1})^+\}. \tag{2}$$

As before let us denote by $W_{n+1,j} (1 \leq j \leq s)$ the quantities on the right side of (2) arranged in ascending order. Then $W_{n+1} = W_{n+1,1}$. The process of interest is thus

$$\mathcal{W}_n = \{W_{n1}, W_{n2}, \ldots, W_{ns}\} \quad (n \geq 0\}. \tag{3}$$

The relation (2) shows that $\{\mathcal{W}_n, n \geq 0\}$ is a time-homogeneous Markov chain on the state space $S = \{x_1, x_2, \ldots, x_n : 0 \leq x_1 \leq x_2 \leq \cdots \leq x_s\}$ (but not a random walk as stated by the authors). Its one-step transition d.f. is given by

$$P(x; y) = \int_0^\infty \int_0^\infty P\{\mathcal{W}_{n+1} \leq y | \mathcal{W}_n = x, v_{n+1} = v, u_{n+1} = u\}\ dA(u)\ dB(v) \tag{4}$$

with $x = (x_1, x_2, \ldots, x_s), y = (y_1, y_2, \ldots, y_s), x \in S, y \in S$. Let F denote the limit d.f. of $\mathcal{W}_n$ as $n \to \infty$. The main result of the first paper is the following:

Theorem 1. If $\rho \geq 1$, then $F(y) = 0$ for all $y \in S$, while if $\rho < 1$, then F is the unique solution of the integral equation

$$F(y) = \int_{x \in S} F(dx)\ P(x; y) \tag{5}$$

such that F is a proper d.f. concentrated on S.

A key result of the second paper is the following:

Theorem 2. Let $\rho < 1$, and for any positive k, $E(v_1^{k+1}) < \infty$. Then with probability one,

$$\frac{1}{N} \sum_{n=1}^{N} (W_{n1})^k \to \int x_1^k F(dx) \quad \text{as} \quad N \to \infty \tag{6}$$

the limit being finite.

The authors use the concept of *stochastic ordering* at several places in their proof of Theorem 1. To convey the essence of their arguments, let Σ denote the given queueing system and Σ' a second s-server system whose inter-arrival times $\{u_k'\}$ and service times $\{v_k'\}$ are such that

$$u_k' \leq u_k \quad \text{and} \quad v_k' \geq v_k \quad (k \geq 1) \tag{7}$$

in distribution (that is, denoting by primes the random variables and d.f.'s in Σ', we have $A'(t) \geq A(t)$ and $B'(t) \leq B(t)$ for $-\infty < t < \infty$). Then one would intuitively expect Σ' to be more congested than Σ in the sense that $W_{n1}' \geq W_{n1}$. This is in fact true for the process (3), as stated in the following.

Theorem 3. For two s-server systems Σ and Σ' described above and satisfying the inequalities (7), let $\mathcal{W}_0' \geq \mathcal{W}_0$. Then

$$\mathcal{W}_n' \geq \mathcal{W}_n \quad (n \geq 0). \tag{8}$$

The authors choose A' and B' to be arithmetic distributions, so that the process $\{\mathcal{W}_n'\}$ is a countable state space Markov chain. They apply the basic theorems of Feller to establish the ergodicity of that chain. It follows from (8) that in the case $\rho < 1$ the limit d.f. F is proper. To deal with the case $\rho \geq 1$ the authors prove the following:

Theorem 4. Let Σ'' be a single server system with inter-arrival times $\{s\ u_k\}$ and service times $\{v_k\}$, and assume that $P\{v_k - s\ u_k = 0\} < 1$. Let W_n'' be the n-th customer's waiting time in Σ'', with $W_0'' \equiv 0$. Then with probability one

$$W_n'' \leq \sum_{j=1}^{s} W_{nj} \quad (n \geq 0). \tag{9}$$

Note that both Σ and Σ'' have the same traffic intensity ρ. If $\rho \geq 1$, then for the limit d.f. F'' of W_n'' we have $F''(y) \equiv 0$. From (9) it follows that $F(y) \equiv 0$.

The arguments sketched above perhaps represent the first use of stochastic ordering to establish convergence of stochastic processes. In any case the results (8)–(9) are certainly the earliest ones on stochastic comparisons of queues. For single server queues the result (8) was proved by Daley and Moran (1968), but they seem to be unaware of this more general result.

Recently there has been renewed interest in multi-server queues. Charlot, Ghidouche and Hamami (1978) have used the *recurrence criteria* developed by T. E. Harris to establish Theorem 1. Whitt (1982) further exploits the concept of stochastic ordering to give a new proof of Theorem 1. He bounds the given process $\mathcal{W}_n$ above and below by sequences of countable state space Markov processes that converge to it. He also gives additional references on this topic.

The integral equation (5) is not of any standard type and the authors do not provide any examples to indicate how it can be solved. This is in contrast with the case $s = 1$, where the transition d.f. $P(x; y) = K(y - x)$, K being the d.f. of $X_k = v_k - u_k$ $(k \geq 1)$, and consequently (1) reduces to an equation of the Wiener-Hopf type and can be solved in some special cases (see Prabhu (1965)).

For $s > 1$ the marginal process $\{W_{n1}\}$ is not Markovian in general. Its limit d.f. F^* is given by

$$F^*(y_1) = \int_{x \in S} F(dx)\, P(x; \bar{y}_1) \quad (y \geq 0) \tag{10}$$

where $\bar{y} = (y_1, \infty, \infty, \ldots, \infty)$. The random variables $(W_{n2}, W_{n2}, \ldots, W_{ns})$ play the role of *supplementary* variables that make the process (3) Markovian. They are also of interst in themselves, since the sum

$$L_n = \sum_{j=1}^{s} W_{nj} \quad (n \geq 0) \tag{11}$$

represents the residual workload of the system at the epoch t_n. The amount of work actually performed by the system is given by $\mathcal{B}_n = (v_1 + v_2 + \cdots + v_n) - L_n$. Further, let $\mathcal{I}_n = (u_1 + u_2 + \cdots + u_n) - \mathcal{B}_n$ in analogy with the single server system where $\mathcal{I}_n$ is the total idle time up to t_n (see Prabhu (1980)). Then we have

$$L_n = S_n + \mathcal{I}_n \quad (n \geq 0), \tag{12}$$

where $S_0 \equiv 0, S_n = S_1 + S_2 + \cdots + S_n$ $(n \geq 1)$.

References

Charlot, F., Ghidouche, M. A. and Hamami, M. (1978). Irreducibility and recurrence in the sense of Harris for the waiting times in the GI/G/q queues. *Z. Wahrsch. Verw. Gebiete.*, 43, 187–203 (in French).

Daley, D. J. and Moran, P. A. P. (1968). Two sided inequalities for waiting time and queue size in GI/G/1. *Theory Prob. Appl.*, 13, 356– 358.

Kendall, D. G. (1951). Some problems in the theory of queues. *J. Roy. Statist. Soc. B*, 13, 151–185.

Kendall, D. G. (1954). Stochastic processes occurring in the theory of queues and their analysis by the method of the imbedded Markov chain. *Ann. Math. Statist.*, 24, 338–354.

Lindley, D. V. (1952). The theory of queues with a single server. *Proc. Camb. Phil. Soc.*, 48, 277–289.

Prabhu, N. U. (1965). *Queues and Inventories.* John Wiley: New York.

Prabhu, N. U. (1980). *Stochastic Storage Processes.* Springer– Verlag: New York.

Smith, W. L. (1953). On the distribution of queueing times. *Proc. Camb. Phil. Soc.*, 49, 449–461.

Whitt, W. (1982). Existence of limiting distributions in the GI/G/s queue. *Math. Oper. Res.*, 7, 88–94.

COMMENTARY ON PAPERS [13], [15], [21], [25], [32]

Ronald Pyke
University of Washington

The beginning of the story about Jack Kiefer's contributions to the asymptotic theory of the empirical distribution function (d.f.) F_n takes place in [13]. The emphasis in this early paper is upon F_n's role as the "observed" in goodness-of-it statistics representable as $\|$"observed"—"expected"$\|$ for some suitable norm or distance $\|\cdot\|$. The main purpose of the paper is to demonstrate a superiority of goodness-of-fit statistics based on F_n, such as those of Kolmogorov, Smirnov, Cramér and von Mises, over the more traditional chi-square statistics. The main hypothesis-testing problem considered is the classically important one of testing the composite $H_0 : F$ is normal versus the full alternative $F_A : F$ is not normal.

This paper contains none of the precise technical estimates and bounds that characterize most of Kiefer's later work on the empirical d.f. Rather, the asymptotic emphasis is on what we would now call the weak convergence of the empirical processes $W_n^F = n^{1/2}(F_n - F)$. A unique aspect of these early results is their treatment of the case in which parameters in the true d.f. F are estimated! In the past decade, some twenty years after [13], the subject of empirical processes with estimated parameters has received considerable attention, and the basic methods remain essentially unchanged.

The more technical approximations pertaining to the asymptotic behavior of F_n begin in [15]. Strikingly enough, the key result of this paper, which initiates an important thread that is traceable through much of Kiefer's subsequent research, is designated as a preliminary lemma, namely Lemma 2. This result provides a powerful exponential bound for the tail probabilities of the Kolmogorov statistic, namely

$$P\{n^{1/2}D_n \geq r\} \leq ce^{-2r^2}, \quad r \geq 0, \tag{1}$$

where $D_n = \|F_n - F\|$, $\|\cdot\|$ is the sup-norm and c is a constant that depends neither on n nor on F.

The proof of (1) is a *tour de force*, providing a clear indication of the authors' outstanding abilities to identify and solve important complex problems which many of us might not dare to hope were even tractable. The bound is actually proved for the tails of the one-sided Kolmogorov-Smirnov statistic D_n^+ which suffices to prove (1), and the proof starts with a known explicit expression for the d.f. of D_n^+. We note that exact expressions are not available in the multidimensional cases discussed below, necessitating the use of different methods. Regardless of methods, however, and many have been tried, the result (1) still remains as one of the more difficult to prove important facts of asymptotic statistics.

Upon noting that $n^{1/2}D_n = \|W_n^F\|$, the reader will observe that (1) provides a uniform bound for the maximum deviation of the empirical process W_n^F and it is in this context that this thread of research has become so important today. In [15], however, this precise estimate was derived as the main tool necesary to establish that F_n is an asymptotically minimax estimator of F under several classes of loss functions. Although essential to the proof of this decision-theoretic result, I believe it is fair to say that the result's provision of simple bounds for percentiles of the statistics D_n and D_n^+ may have at first been viewed by many as its most important role.

In this latter spirit of providing probability bounds for Kolmogorov- Smirnov type statistics, the paper [21] extends the exponential bound of (1) to the case in which the data $X_1, \cdots, X_n$ are m-dimensional random vectors. This bound, which in the one-dimensional case of [15] was but a lemma, now becomes the main result (Theorem 1):

$$(2) \qquad P\{n^{1/2}D_n \geq r\} \leq c_o exp(-cr^2), \quad r \geq 0,$$

where the constants c_o and c depend on the dimension m but not on n nor F. Although two pages beginning in the middle of page 181 were devoted by the authors to the determination of a suitable constant c, the main interest was clearly not in obtaining the best bounds but rather in providing a sufficiently good bound for use in establishing the asymptotically minimax property for F_n in the vector case. The bound (2) was in fact used for this purpose in [25]. Although labelled as the main theorem, it is clear that the authors viewed (2) primarily as a preliminary result for [25].

As a second result in [25], the authors shows that the d.f.'s of the multi- dimensional Kolmogorov-Smirov statistics, both one-and two-sided, converge in law, and implicitly characterize the limiting distributions as those of the corresponding supremum of a process that is known today as a tied-down Brownian sheet, say W^F. This is a mean zero, Gaussian process with

$$\operatorname{cov}(W^F(s), W^F(t)) = F(s \wedge t) - F(s)F(t)$$

in which s and t are points in the m-dimensional unit cube $[0,1]^m$, $\wedge$ is interpreted coordinatewise and F is transformed, without loss of generality, to have its support in $[0,1]^m$. The last sentence on the first page of [25] states that these limiting distributions "for $m > 1$ are at present writing unknown". This is still true 25 years later.

When one turns to the next paper in this sequence, [32], one sees that the emphasis now has clearly been placed upon the bound itself and a search for the optimal constant c. The applications are no longer confined to limiting distributions for statistics but include almost sure results such as the law of the iterated logarithm for the m-dimensional Kolmogorov-Smirov statistics. The main result in [32] is the improvement of (2) to: For all $\varepsilon > 0$ and integers $m \geqq 1$,

$$(3) \qquad P\{n^{1/2} D_n \geq r\} \leq c(\varepsilon, m) e^{-(2-\varepsilon)r^2}, \quad r \geq 0,$$

where $c(\varepsilon, m)$ is a constant independent of F. When one considers that in [25] the best constants in the exponent of (2) converged to 0 as $m \to \infty$ and that for $m = 3, c = .000107$, then one can see the degree of improvement contained in (3). Although all three results, (1), (2) and (3), are proved using different methods, they all involve precise estimates on the tails of normalized binomial distributions. Today, the 1962 inequality of Bennett (1962) which followed [32] by one year, is usually applied for these purposes, but the reader should be interested in the inequality stated as Lemma 1 of [32].

In all of these results, the motivation for seeking "2" as the optimal constant in the bounds exponent derives from the fact that when $F(x) = 1/2$, $W_n^F(x)$ is a normalized binomial $B(n, 1/2)$ random variable. Therefore, by the classical central limit theorem, its d.f. can be bounded in the tails by $\exp(-2r^2)$, since, for this case, its variance is $p(1-p) = F(x)\{1 -$

$F(x)\} = 1/4$. Moreover, for any other value of $p = F(x)$, $p(1-p) \leq 1/4$. Thus for each fixed $x, c = 2$ is the largest constant bound; the challenge is in finding the largest constant that holds uniformly in x as required for D_n. By a simple example Kiefer shows in [32] that the constant $c = 2$ can not be achieved when $m > 1$, his result (3) is the best possible of this form.

As I have mentioned, the details in these papers provide ample evidence of exceptional ingenuity and skill with complicated technical and formulational problems. However, rather than commenting on these details, let me instead draw attention to two conceptual gems that stand out so brilliantly to me as I read these papers again more than two decades after they first appeared.

Notice first the following statement from page 174 of [25],

> "We also point out that the supremum operation can be performed over a larger class of sets without affecting the result."

The "larger classes of sets" that the authors had in mind were small by today's standards, as indicated for example on page 183 of [25],

> "It is obvious that Theorem 1 (essentially (2) above) applies also to the case when the supremum is taken over any of several larger classes of sets such as, for example, that which consists of all rectangular parallelpipeds with sides parallel to the coordinate planes. This will be of interest in statistical applications... ."

Despite this relative "smallness", their recognition of the probabilistic naturalness and statistical importance of viewing empirical processes as set-indexed processes is clear. It was in the following year, that Kuiper (1960) proposed for $m = 1$, in the context of circular data, a Kolmogorov-type statistic defined as the supremum over all intervals on the line. For related references, see Brunk (1962), in which Kolmogorov-type statistics determined by general classes of sets are also defined. (The study of large deviations of empirical processes indexed by m-dimensional rectangles is still receiving attention today, even when the study of empirical processes centers primarily on much larger index families of sets.) The last sentence [25] also refers to the extendability of the limiting distribution result of Theorem 2 to these cases of "larger classes of sets". Here one might note the above papers of Kuiper and Brunk in which the explicit limiting d.f. of Kuiper's statistic is obtained. No such exact results for $m > 1$ are known.

The second concept in these papers that shows considerable foresightedness is that of stochastic processes indexed by "m-dimensional time". For example, on page 175 of [25] discussion is given of the derivation of the limiting d.f.'s on m-dimensional Kolmogorov-Smirnov statistics "by consideration of a Gaussian process (depending on F) with m-dimensinal time." This is the tied-down Brownian Sheet described above. The phrase "m-dimensional time" is also mentioned three times in [32], on page 651. Although processes indexed by multi-dimensional parameters had been studied prior to these papers — for example, Lévy's isotropic generalization of Brownian Motion to m-dimensional time which began with Lévy (1940), and the introduction of Brownian Sheet by Chentsov (1956) – -their inclusion in this more statistical context shows the type of foresightedness possessed by Kiefer and his coauthors. Particularly striking to me is the definition in [32] of a partial-sum process generated by an m-dimensional array of independent random variables, as well as his mention of "semi-martigales with m-dimensional time". Moreover, the standard Brownian Sheet is given as an example of a process indexed by m- dimensional *continuous* time. The discussion here of these concepts precede by several years the intensive activity that is presently associated with partial-sum processes and martingales indexed by multi- dimensional time. (For the former, see Pyke (1983), Bass and Pyke (1984) and references contained therein; for the latter see the collection of papers edited by Kőrezlioğlu, Mazziotto and Szpirglas (1981).) The Lévy-type inequalities for symmetric partial-sum arrays and Brownian Sheet given on page 651 of [32] have not been adequately justified here; for related inequalities, see Wichura (1969) and Klesov (1982).

In summary, the papers reviewed above initiate and significantly develop the theory of asymptotic bounds for empirical processes, and make important early contributions to the weak convergence of empirical processes when $m \geq 1$. Both of these topics today represent research areas of considerable activity and importance. For current information on exponential bounds, see Alexander (1983) and the references contained therein. For a current presentation of weak convergence results for empirical processes, see Gaenssler (1984) and the references contained therein.

References

Alexander, K. S. (1983). Probability inequalities for empirical processes and a law of the iterated logarithm. (Unpublished)

Bass, R. F. and Pyke, R. (1984). Functional law of the iterated logarithm and uniform central limit theorem for partial-sum processes indexed by sets. *Ann. Prob.*, 12, 13–34.

Bennett, G. (1962). Probability inequalities for the sum of independent random variables. *J. Amer. Statist. Assoc.*, 57, 33–45.

Brunk, H. D. (1962). On the range of the difference between hypothetical distribution function and Pyke's modified empirical distribution function. *Ann. Math. Statist.*, 33, 525–532.

Chentsov, N. N. (1956). Wiener random fields depending on several parameters. *Dokl. Akad. Nauk SSR (N.S.)*, 106, 607–609.

Gaenssler, P. (1984). Empirical Processes. *IMS Lecture Notes- Monograph Series*, No. 3. Hayward, CA: Institute of Mathematical Statistics.

Klesov, O. I. (1982). The law of the iterated logarithm for multiple sums. *Teor. Veroyatnost. i Mat. Statist.*, 27, 60–67. (Translated in *Theory Probab. and Math. Statist.* 27 (1983) 65–72.)

Kőrezlioğlu, H., Mazziotto, G. and Szpirglas, J. (1981). Processus aléatoires á deux indices. Lectur Notes in Mathematics; 863, Springer-Verlag: Berlin.

Kuiper, N. H. (1960). Tests concerning random points on a circle. *Nederl. Akad. Wetensch. Proc. Ser. A., Indag. Math.*, 22, 38–47.

Lévy, P. (1940). Le mouvement Brownien plan. *Amer. J. Math.*, 62, 487–550.

Pyke, R. (1973). Partial sums of matrix arrays, and Brownian sheets. *Stochastic Analysis*, (ed. D. G. Kendall and E. F. Harding), 331–348, Wiley: London.

Pyke, R. (1983). A uniform Central Limit Theorem for partial-sum processes indexed by sets. *Probability, Statistics, and Analysis*, (ed. J. F. C. Kingman and G. E. H. Reuter), 219–240. Cambridge University Press: Cambridge.

Wichura, M. J. (1969). Inequalities with application to the weak convergence of random processes with multidimensional time parameter. *Ann. Math. Statist.*, 40, 681–687.

COMMENTARY ON PAPERS [50], [51], [56]

P. Révész

Mathematical Institute, Budapest, Hungary

The theory of the strong invariance principle was initiated by the paper of Strassen (1964). He proposed to construct a Wiener process $W(t)$ and a sequence $\{X_i\}$ of i.i.d. random variables with a given distribution function F on a probability space in such a way that $|S_n - W(n)|$ $(S_n = \sum_{i=1}^n x_i)$ would be small in the sense that the relation $\lim_{n\to\infty}(g(n))^{-1}|S_n - W(n)| = 0$ a.s. should hold for a suitably increasing function g. Using the Skorokhod embedding scheme, Strassen himself (1965) proved that, assuming $EX_1 = 0,\ EX_1^2 = 1,\ EX_1^4 < \infty$ and choosing $g_0(n) = (n \text{ loglog } n)^{\frac{1}{4}}(\log n)^{\frac{1}{2}}$ one can construct W and $\{X_i\}$ such a way that $(g_0(n))^{-1}|S_n - W(n)| = O(1)$ a.s.

Already in his 1965 paper Strassen asked: can the Skorohod embedding or any other construction produce a result saying that $(g_0(n))^{-1}|S_n - W(n)| = o(1)$ a.s. Similar questions were also formulated by Breiman (1967) and Borovkov (1973).

Paper [50] gives a negative answer to this question, proving that using the Skorokhod embedding scheme, $\limsup_{n\to\infty}(g_0(n))^{-1}|S_n - W(n)| = (2\beta)^{\frac{1}{2}}$, where β is a constant depending on F, and $\beta = 0$ if and only if F is $\mathcal{N}(0,1)$.

Later on it turned out that making use of a different construction a much better rate can be obtained (cf. Komlós-Major-Tusnády, (1976)).

The importance of such a result is shown by the fact that having a statement like $(g(n))^{-1}(S_n - W(n)) = o(1)$ a.s. with $g(n) = o(n^{\frac{1}{2}})$, and knowing practically any strong result (such as the law of the iterated logarithm) for W, one automatically get the very same result for S_n.

In the light of Strassen's strong invariance principle, it was only natural to look for analogous approximations also for the empirical process is the empirical distribution function of a sequence of independent, uniform $(0,1)$ random variables $U_1, U_2, \ldots, U_n$. The first such result was obtained by Breiman (1968) and Brillinger (1969) who proved that one can construct a sequence $\{B_n(x)\}$ of Brownian bridges such that $\sup_{0\le x\le 1}|\alpha_n(x)-B_n(x)| = O(n^{-\frac{1}{4}}(\log n)^{\frac{1}{2}}(\log\log n)^{\frac{1}{4}})$ a.s. should hold. However this result is not really a strong approximation result in the sense that Strassen's result is. In fact Brillinger's result gives an approximation for each n fixed, and it does not say anything about the possible behavior of the sequence $\{B_n\}$ in n. Consequently no strong law type behavior of $\alpha_n(x)$ (as, for example, Chung's law of iterated logarithm (1949)) can be deduced from it.

Paper [51] was the first one to call attention to the desirability of viewing the empirical process $\alpha_n(x)$ as a two parameter process and of finding a strong approximation theorem for $\alpha_n(x)$ in terms of an appropriate two dimensional Gaussian process. As a solution of this problem in [56] he defined a Gaussian process $\{K(x,t); 0\le x\le 1,\quad 0\le t\le\infty\}$ (called the Kiefer process today) and proved that on a suitable probability space one can define the sequence $\{U_n\}$ and the process $K(x,t)$ such a say that $\sup_{0\le x\le 1}|n^{\frac{1}{2}}\alpha_n(x) - K(x,n)| = O(n^{\frac{1}{3}}(\log n)^{\frac{2}{3}})$ should hold. This result clearly implies among others that Chung's law of iterated logarithm (proved for α_n) is inherited by K.

Later on it turned out that the rate of convergence in the above mentioned theorem of Kiefer can be improved (cf. Komlós-Major-Tusnády (1975)) and that it can be extended to the multivariate case (cf. Borisov (1982)).

References

Borisov, S. I. (1982). An approximation of empirical fields, in *Nonparametric Statistical Inference*. Budapest, Hungary: Janos Bolyai; New York: North-Holland Publishing Co.

Borovkov, A. A. (1973). On the speed of convergence in the invariance principle. *Theory of Probability and Its Applications*, 18.

Breiman, L. (1967). On the tail behavior of sums of independent random variables. *Z. Wahrsch. Verw. Gebiete*, 9, 20–25.

Brillinger, P. L. (1969). An asymptotic representation of the sample distribution function.

Bull. Amer. Math. Soc., 75, 545–547.

Chung, K. L. (1949). An estimate concerning the Kolmogorov limit distribution. *Trans. Amer. Math. Soc.*, 64, 205–233.

Komlós, J., Major, P., Tusnády, G. (1975–76). An approximation of partial sums of independent R.V.'s and the sample D.F. I-II. *Z. Wahrsch. Verw. Gebiete*, 32, 111–131, 37, 33–58.

Strassen, V. (1964). An invariance principle for the law of iterated logarithm. *Z. Wahrsch. Verw. Gebiete*, 3, 211–226.

Strassen, V. (1965). Almost sure behaviour of sums of independent random variables and martingales. *Proc. Fifth Berkeley Symp. Math. Statist. and Probability*, 2, 315–344.

COMMENTARY ON PAPERS [47], [52], [54]

P. Révész

Mathematical Institute, Budapest, Hungary

Let $X_1, X_2, \ldots$ be i.i.d. random variables with distribution function F and density function $f = F'$. Let F_n be the empirical distribution based on the sample $X_1, X_2, \ldots, X_n$, and consider the empirical process

$$\beta_n(x) = n^{\frac{1}{2}}(F_n(x) - F(x))$$

and the quantile process

$$q_n(x) = n^{\frac{1}{2}} f(F^{-1}(x))(F_n^{-1}(x) - F(x)).$$

Bahadur (1966) observed that the distance $R_n(x) = q_n(x) + \beta_n(F^{-1}(x))$ between $q_n(x)$ and $-\beta_n(F^{-1}(x))$ is very small. In fact, under some regularity conditions he proved that $R_n(x) = O(n^{-\frac{1}{4}}(\log n)^{\frac{1}{2}}(\log\log n)^{\frac{1}{4}}$ a.s. Kiefer succeeded in giving an exact description of the behaviour of $R_n(x)$. The main result of [47] says that $\limsup_{n\to\infty} n^{\frac{1}{4}}(\log\log n)^{-\frac{3}{4}} R_n(x) = 2^{\frac{5}{4}} \cdot 3^{\frac{3}{4}}(x(1-x))^{\frac{1}{4}}$ a.s. In [52] he reconsidered this problem and characterized the limit behaviour of $\sup R_n(x)$, proving a limit distribution theorem as well as a strong theorem.

Theorems stating the smallness of R_n are very useful when we have a weak or strong limit theorem on β_n and we want to prove the same theorem for q_n or vice versa. Also strong invariance principle type results can be obtained for q_n having such an invariance result for β_n and knowing the smallness of R_n. Shorack (1982) showed that the opposite direction can be also followed. Namely, one can prove results for R_n starting first with invariance type results.

The above mentioned results help us to obtain a *global* description of q_n. However, a different question is to study the behavior of $q_n(p_n)$ when $p_n \downarrow 0$. This question was studied and fully solved in [54], obtaining a complete characterization of the lower and upper classes of the sequence $\{q_n(p_n)\}$. In the case of uniform $[0,1]$ random variables the investigation of the properties of $\{\beta_n(p_n)\}$ is closely related and the same results were obtained. This type of results suggests investigating the limit behavior of the sequence $\{\sup_{p_n \le x \le 1-p_n} (x(1-x))^{-\frac{1}{2}} |\beta_n(x)|\}$. This problem was studied and solved by Csáki (1977).

References

Bahadur, R. R. (1966). A note on quantiles in large samples. *Ann. Math. Statist.*, 37, 577–580.

Csáki, E. (1977). The law of iterated logarithm for normalized empirical distribution function. *Z. Wahrsch. Verw. Gebiete*, 38, 147–167.

Shorack, G. R. (1982). Kiefer's theorem via the Hungarian construction. *Z. Wahrsch. Verw. Gebiete*, 61, 369–373.

COMMENTARY ON PAPERS [13], [15], [25], [65], [68]

P. Warwick Millar
University of California, Berkeley

The fundamental paper here is [15] – a milestone in the develpoment of the theory of asymptotically optimal non-parametric inference. The key insight was that the least favorable distributions at time n were concentrated on neighborhoods of measures within $cn^{-\frac{1}{2}}$ of a fixed measure F_0, distance measured by (for example) the Kolmogorov-Smirnov metric; that is, the crux of the asymptotic minimax argument was *local*. In [15], F_0 was the uniform distribution and the "neighborhoods" were constructed by perturbations of F_0 as described in (3.3).

This observation that the local behavior was crucial has spawned an extensive development, leading to the fundamental concept of local asymptotic minimaxity (LAM). A high point in this development is (Hájek, 1972), wherein the author gave a general LAM theorem for parametric models, a development foreshadowed by [15]; Hájek's development connects the LAM property with a notion of efficiency that makes rigorous the classical notion of Fisher. LeCam (1972) abstracted the LAM concept further, extending it beyond Hájek's parametric formulation and connecting it to his (LeCam's) notion of convergence of statistical experiments. By extending the type of neighborhoods, Beran (1981), Millar (1981), Rieder (1981), for example, gave developments of robustness based precisely on notions of stability implied by a LAM structure. Recent expositions of these developments can be found in Ibragimov-Hasminskii (1981) and Millar (1983).

The heuristic suggestions of Section 7 of [15] have not been pursued in the form given there. However, Levit (1978) and Millar (1979), basing developments respectively on Hájek's and LeCam's general asymptotic minimax theorems, have given purely structural proofs of the main results of [15], side-stepping the difficult calculations of the approach used by Dvoretzky, Kiefer, and Wolfowitz.

Papers [65] and [68] open a new problem: if C is a collection of c.d.f.'s, when is the empirical c.d.f., $\hat{F}_n$, an asymptotically minimax estimate of the c.d.f. F, when it is known a priori that F belongs to C. As before, the problem is essentially local. In [65], [68] Kiefer and Wolfowitz show that $\hat{F}_n$ is asymptotically minimax when C is the class of convex c.d.f.'s. Millar (1979) gave abstract, non-computational proofs of this, together with a discussion of a number of other classes C (for example all c.d.f.'s with decreasing failure rate). Millar also shows that, if C is "too small", then $\hat{F}_n$ cannot be asymptotically minimax. Papers [69] and [70] go much further, showing in the special case where C is the class of convex c.d.f's, how $\hat{F}_n$ may be replaced by an estimator $\hat{C}_n \in C$ which does not lose efficiency (for the interesting classes C, $\hat{F}_n$ will fail to belong to C with probability 1). Wang (1982) has recently shown how $\hat{F}_n$ can be projected to the class of c.d.f.'s having a density symmetric about some point by intricate calculations as in [65]. It is clear that the class of decreasing failure rate distributions can also be treated; Lo (1981) treats the analogous problem for the class of c.d.f.'s with density symmetric about some unknown point. As of this writing, however, there is no clear structural development (involving, perhaps, conditions on the geometric 'shape' of C) which gives general guidelines as to when $\hat{F}_n$, when it is asymptotically minimax relative to C, can be replaced by an element of C without losing efficiency.

References

Beran, R. J. (1981). Efficient robust tests in parametric models. *Z. Wahrsch. Verw. Gebiete*, 57, 73–86.

Hájek, J. (1972). Local asymptotic minimax and admissibility in estimation. *Proc. Sixth Berkeley Symp. Math. Statist. and Probability*, 1, 175–194.

Hasminskii, R. Z. and Ibragimov, I. A. (1981). *Statistical Estimation*. Springer-Verlag: New York.

LeCam, L. (1972). Limits of experiments. *Proc. Sixth Berkeley Symp. Math. Statist. and Probability*, 1, 245–261.

Levit, B. Ya. (1978). Infinite dimensional information inequalities. *Theory Probab. Appl.*, 23,

371–377.

Lo, S.-H (1981). Locally asymptotic minimax estimation for symmetric distributions functions and shift parameters. Doctoral dissertation, University of California: Berkeley.

Millar, P. W. (1979). Asymptotic minimax theorems for the sample distribution function. *Z. Warhsch. Verw. Gebiete*, 48, 233–252.

Millar, P. W. (1981). Robust estimation via minimum distance methods. *Z. Wahrsch. Verw. Gebiete*, 55, 73–89.

Millar, P. W. (1983). The minimax principle in asymptotic statistical theory in *Lecture Notes in Mathematics*, Vol. 976, (ed. by P. L. Hennequin) Ecole d'Eté de Probabilités de Sain-Flour X1-1981, Springer-Verlag: New York.

Rieder, H. (1981). On local asymptotic minimaxity and admissibility in robust estimation. *Ann. Statist.*, 9, 266–277.

Wang, J.-L. (1982). Asymptotically minimax estimators for distributions with increasing failure rate. Doctoral dissertation, University of California: Berkeley.

COMMENTARY ON PAPERS [67], [69], [70], [71]

James Berger

Purdue University

1. Introduction.

Among the earliest criticisms directed towards the frequentist school of statistics were criticisms based on conditioning concepts. As anomalies based on conditioning became more apparent, frequentist statisticians tended to react in one of two ways. Some rejected the frequentist approach, typically adopting some type of Bayesian viewpoint. Others simply ignored the difficulties, perhaps from a feeling that they were not of practical importance. Kiefer took neither route. His devotion to scientific truth and fundamental understanding would not allow him to ignore the problem, but his strong belief in the necessity for frequentist justification prevented the escape into Bayesianism. The result was a massive undertaking of retooling frequentist theory to accommodate conditional needs. His success in this retooling was remarkable. A brief review of the central ideas in Kiefer's development is presented in Sections 2 through 6, followed by an overview of the approach in Section 7.

It is worthwhile to begin with three examples to indicate the conditioning problem, and also to use in illustrations of Kiefer's approach. We use Kiefer's notation of observing a random quantity X having probability distribution $P_\omega, \omega \in \Omega$ unknown. Kiefer essentially restricted himself to consideration of problems where the "decision rule" δ resulted in either a correct or incorrect conclusion (zero-one loss), so only such examples will be considered.

Example 1. Suppose $X = (X_1, X_2)$ where the X_i are independent with

$$P_\omega\{X_i = \omega - 1\} = P_\omega\{X_i = \omega + 1\} = 1/2.$$

Consider the confidence procedure

$$\delta(X) = \begin{cases} \text{the point } (X_1 + X_2)/2, & \text{if } |X_1 - X_2| = 2, \\ \text{the point } (X_1 - 1), & \text{if } |X_1 - X_2| = 0. \end{cases}$$

An easy calculation shows that

$$P_\omega\{\delta(X) \text{ contains } \omega\} = .75 \text{ for all } \omega.$$

Hence one can, with complete frequentist validity, report $\delta(X)$ and 75% "confidence." This seems intuitively ridiculous, however, since, when $|X_1 - X_2| = 2$, one *knows* that $\delta(x) = \omega$, whereas, when $|X_1 - X_2| = 0, \omega$ is either $X_1 - 1$ or $X_1 + 1$ with the data giving no preference to either possible value. Thus a sensible analysis would be to report 100% "confidence" when $|X_1 - X_2| = 2$ and 50% "confidence" when $|X_1 - X_2| = 0$.

Example 2. Let $X \sim \mathcal{N}(\omega, 1)$, and suppose it is desired to test $H_0 : \omega \leq -1$ versus $H_a : \omega \geq 1$. Let $\delta(X)$ be the test which rejects when $X \geq 0$. Then

$$P_\omega\{\delta(X) \text{ makes the correct decision}\} = \Phi(|\omega|),$$

where Φ is the standard normal c.d.f. Since

$$\sup_{|\omega|\geq 1} \{1 - \Phi(|\omega|)\} = 0.16,$$

0.16 is the error probability that would commonly be reported, and this would be the report regardless of the observation X. But, intuitively, $X = 0$ does not provide any evidence to distinguish between the two hypotheses, while $X = 10$ provides overwhelming evidence in favor of H_a. The frequentist report seems a severe underestimate of error in the first case, and a severe overestimate in the second.

Example 3. Consider the situation of Example 2, but let the hypotheses be $H_0 : \omega \leq 0$ versus $H_a : \omega > 0$. Now

$$\sup_{\omega}\{1 - \Phi(|\omega|)\} = 1/2,$$

so no useful frequentist bound on error probability can be given. The observation of $X = 10$ still seems overwhelming evidence in favor of H_a, however.

2. Conditional Confidence.

The bulk of Kiefer's development deals with what he called conditional confidence. The most readable account is in [69], while the most mathematically rigorous exposition is in [67]. C. Brownie also played a role in this development (see [70]).

The basic idea of conditional confidence is to partition the sample space $\mathcal{X}$ into disjoint sets $\{C^b, b \in B\}$, and then evaluate a procedure conditionally on these sets, in the usual frequentist manner. Thus, of relevance would be

$$\Gamma_\omega^b = P_\omega\{\delta(X) \text{ makes the correct decision } |C^b\}.$$

Defining $\Gamma_\omega(X)$ to be that Γ_ω^b for which $X \in C^b$, the "report" from an experiment would be $(\delta(X), \Gamma_\omega(X))$. This has the frequentist justification that, on the average over all times that X is in $C^b, \delta(X)$ will make the correct decision Γ_ω^b proportion of the time.

Example 1 (continued). Define

$$C^0 = \{X_1 - X_1| = 2\} \text{ and } C^1 = \{X : |X_1 - X_2| = 0\}.$$

Then

$$\Gamma_\omega^0 = P_\omega\{\delta(X) \text{ contains } \omega|C^0\} = 1,$$

$$\Gamma_\omega^1 = P_\omega\{\delta(X) \text{ contains } \omega|C^1\} = 1/2,$$

and so the conditional reports coincide with common sense and have frequentist validity.

Example 2 (continued). Let $B = (0, \infty)$, and define $C^b = \{b, -b\}$ (just two points). Then, letting $f(x|\omega)$ denote the normal density,

$$\begin{aligned}\Gamma_\omega^b &= P_\omega\{\delta(X) \text{ makes the correct decision } |C^b\} \\ &= 1 - \frac{f(-(\text{sgn}\omega)b|\omega)}{f(-b|\omega) + f(b|\omega)} \\ &= \frac{1}{1 + \exp\{-2b|\omega|\}}.\end{aligned}$$

Upon observing X, one thus reports the decision $\delta(X)$ and $\Gamma_\omega(X) = [1 + \exp\{-2|X\omega|\}]^{-1}$. Note that

$$\inf_\omega \Gamma_\omega(X) = [1 + \exp\{-2|X|\}]^{-1},$$

so that, when X is near zero, the conditional confidence (one minus the error probability) reported will be near $1/2$, while, when X is far from zero, the conditional confidence reported will be very high. This corresponds to intuition.

Example 3 (continued). Here $\Gamma_\omega(X)$ will be as in Example 2, but now

$$\inf_\omega \Gamma_\omega(X) = \Gamma_0(X) = 1/2,$$

a useless bound. Furthermore, it can be shown that *any* partition $\{C^b\}$ results in $\inf_\omega \Gamma_\omega(X) \leq 1/2$, so that conditional confidence will fail to remedy the problem of not being able to conclusively justify rejecting when $X = 10$ is observed.

3. Estimated Confidence.

Kiefer's estimated confidence thoery is presented in [69] and [71]. The simplest situation is that in which one can find an unbiased estimator of $\Gamma_\omega = P_\omega\{\delta(X)$ makes the correct decision$\}$. Thus it is desired to find a $\hat{\Gamma}(X)$ such that

$$E_\omega \hat{\Gamma}(X) = \Gamma_\omega \text{ for all } \omega.$$

If such a $\hat{\Gamma}$ can be found, then the report of $\delta(X)$ and $\hat{\Gamma}(X)$ has complete frequentist validity, in the sense that the average proportion of correct decisions in repeated use will converge to the average of the reported confidences $\hat{\Gamma}$.

Example 1 (continued). Clearly an unbiased estimator of $\Gamma_\omega = 0.75$ is

$$\hat{\Gamma}(X) = \begin{cases} 1, & \text{if } |X_1 - X_2| = 2, \\ 1/2, & \text{if } |X_1 - X_2| = 0, \end{cases}$$

and this is a good intuitive report.

It will not always be the case that an unbiased etimator of Γ_ω can be found. Generally, however, one can find a conservatively biased estimator $\hat{\Gamma}(X)$, i.e., an estimator satisfying

$$E\hat{\Gamma}(X) \leq \Gamma_\omega \text{ for all } \omega.$$

Reporting $\hat{\Gamma}(X)$ is then justifiable, in the usual frequentist sense that the average reported performance will be no better than the actual average performance.

Example 3 (continued). No unbiased estimator of $\Gamma_\omega = 1 - \Phi(|\omega|)$ exists, but conservatively biased estimators $\hat{\Gamma}(X)$ can be found which are increasing in $|X|$. Use of such $\hat{\Gamma}$ would allow one to conclusively reject when $X = 10$ is observed. (There is, however, one fly in the ointment: any such $\hat{\Gamma}$ must be smaller than $1/2$ over some region, in order that $E_0\hat{\Gamma}(X) \leq \Gamma_0 = 1/2$. One would feel slightly foolish making a decision between two hypotheses and stating an estimated confidence less than $1/2$, a confidence which could be attained merely by random guessing.)

Kiefer also considers combining conditional and estimated confidence, by finding (conservatively biased) estimators, $\hat{\Gamma}^b(X)$, for Γ^b_ω. Example 6(b) in [69] is one instance of such a combined approach.

4. Choosing Among Conditional Procedures.

There is a huge latitude in choice among conditional confidence, estimated confidence, and combination procedures. Kiefer has relatively little to say about choosing among estimated confidence procedures. This could have simply been due to the fact that estimated confidence theory requires a quite different development than conditional confidence theory, and the latter seems less "radical" to frequentists. Kiefer clearly found estimated confidence theory very attractive, however, partly because it provided a "report" $\hat{\Gamma}(X)$ that was independent of the unknown ω, and partly because it was clearly necessary, in order to deal with problems like Example 3. (Some further thoughts on estimated confidence can be found in Berger (1983)).

As to choice among partitions $\{C^b\}$ in conditional confidence theory. Kiefer had a great deal to say. Two primary guidelines (c.f. Sections 3.3 and 3.5 in [69] and Section 3.3 in [70]) were:

(i) Attempt to make the Γ^b_ω nearly constant in ω, so that one can report a good conditional confidence bound;

(ii) Try to make Γ^b_ω highly variable over b, at least when highly variable measures of conclusiveness seem desirable (as in Examples 1, 2, and 3).

Indeed, for the latter reason, Kiefer leaned towards supporting the "finest" partition possible, subject to compatibility with (i) (c.f., Kiefer's reply to the discussion in [69]). This finest partition he often called the "continuum" partition. The one qualification Kiefer gave, was that, when model robustness is a serious concern, the finer partitions will tend to give less robust answers (c.f. Section 7.3 in [69]).

Another concept Kiefer introduced was that of a monotone partition ([70]; see also Brown (1978)), which has a number of desirable properties. Kiefer also commented that conditioning on partitions defined by ancillary statistics or maximal invariants (as is $T = |X_1 - X_2|$ in Example 1) is natural when they exist.

In general, even with the above aids there will often be an embarassing plethora of conditional procedures from amongst which to choose. Kiefer admitted the problem ([69]), but felt that the arbitrariness of choice of a conditional procedure was unavoidable, and no worse than, say, the Bayesian need to specify a prior distribution. There is, however, the tantalizing suggestion, in his response to Buehler in [69], that one should perhaps try to select a conditional procedure that mimics, as closely as possible, the conditional report from a Bayesian analysis of the problem.

5. Admissibility for Conditional Confidence Theory.

Decision theory always being dear to his heart, Kiefer spent considerable effort in an attempt to set up a formal decision theoretic structure for conditional confidence. It was surely also his hope that many of the possible partitions could be eliminated as inadmissible. He concentrated on looking at

$$G_\omega(t) = P_\omega\{\Gamma_\omega(X) > t, \text{ given that } \delta(X) \text{ makes the correct decision}\}$$

and $P_\omega\{Q_\omega\}$, where

$$Q_\omega = \{\text{the union of all } C^b \text{ such that, on all of } C^b, \delta \text{ makes an incorrect decision for } \omega\}.$$

One would intuitively like G_ω to be large and $P_\omega\{Q_\omega\}$ to be small, and admissibility and inadmissibility was defined by Kiefer, in the obvious manner, based on *both* criteria simultaneously. (Use of either in isolation gives silly answers.)

Complete characterization of admissible and inadmissible procedures was given in [70], for testing simple versus simple hypotheses. The somewhat unfortunate result (see Theorem 3.1) was that, as long as δ is an admissible Neyman-Pearson unconditional test, *any* partition results in an admissible conditional confidence procedure. Thus even the useless partition $\{C^x : x \in \mathcal{X}\}$, for which $\Gamma_\omega(X)$ is one if $\delta(X)$ is correct for ω and 0 otherwise, is admissible.

In [70], Kiefer gives a general treatment of admissibility, though mainly limited to k-hypotheses problems. Not everything is admissible here, but enough procedures are admissible that one again wonders if Kiefer's criteria are not too weak. This feeling is reinforced by the admissibility, in general, of the useless partition mentioned above, as well as the null partition (i.e., unconditional analysis), which in situations such as Example 1 seems definitely undesirable.

More stringent admissibility criteria are mentioned in [67]. Also, Brown (1978) develops quite stringent and appealing criteria, based on $\inf_\omega \Gamma_\omega$, for conditional confidence. Admissibility criteria for estimated confidence are suggested in Berger (1983).

6. Statistical Problems Considered in the Papers.

In the series of papers, several types of statistical problems were considered. The most studied problem was the testing of simple versus simple hypotheses ([70] and [69]). Partitions considered included the "finest" continuum partition and the C^2 partition (which is a partition of $\mathcal{X}$ into two sets, designed to reflect "weak" and "strong" confidence). It was shown how to satisfy (or approximately satisfy) the criteria (i) and (ii) (discussed in Section 4) for these situations.

In [69] and especially in [71], ranking and selection problems were considered, of both Bechhofer's indifference zone variety and Gupta's subset selection type. Analysis was mainly restricted to the selection of the largest mean from 2 or 3 normal populations, although a binomial example was done in [71] to indicate problems with discrete distributions.

In regard to discrete distributions, randomization is sometimes needed to achieve certain conditional confidence characteristics (or to provide a convex class of procedures for decision theoretic analysis). Discussion of randomization is given in Section 3.4 of [70] and in [67].

Confidence sets are discussed in Section 6 of [69], with a variety of interesting examples concerning conditional and estimated confidence being presented. Also, various auxilliary criteria, such as length, are considered.

Sequential analysis is discussed in Section 7.1 of [69]. Of considerable interest here is the possibility (see Example 10) of defining the conditioning sets, C^b, to (at least approximately) be sets such that the stopping time is the same for all $x \in C^b$. The stopping rule is then not involved in the conditional frequentist analysis, a vast simplification (and a desirable one from most non-frequentist conditional perspectives — c.f., Berger (1983)). Note that frequentist validity is still being maintained; this possible combination of frequentist validity and massive simplification opens exciting new horizons in frequentist sequential analysis.

Robustness issues are considered in Section 7.3 of [69], and a nonparametric example of conditional confidence, involving rank order statistics, is given in [71]. Finally, extensions of the conditional confidence concept to general decision theoretic problems is discussed in Section 7.4 [69].

7. An Overview.

There was, of course, a quite large literature, related to conditional frequentist analysis, that preceded Kiefer's work on the subject. His was the first, however, to formalize the ideas to the point of achieving frequentist validity. A key point here (see the discussion by Brown in [69]) is that one must specify (for conditional confidence) an entire partition of $\mathcal{X}$ ahead of time, not just a "relevant subset" to be conditioned on, if it obtains, with reversion to an unconditional confidence statement otherwise. Similarly, the idea of estimated confidence or estimated risk preceded Kiefer, but he appears to have been the first to realize that one could report such estimated confidence with exactly the same frequentist validity as an unconditional report (see also Berger (1983)). References to many of these earlier conditional frequency works can be found in [69].

In Section 3.2 of [69] and Section 3.2 of [70], the relationship of the conditional theories to "enlarged" decision spaces (e.g., "weak acceptance", "strong acceptance", etc., with associated losses) is discussed. There is a substantial difference between the approaches.

Finally, we come to the crucial issue, the relationship of the conditional frequentist approaches to other conditional approaches, such as the Bayesian approach. One might hope that the two could be reconciled. Such appears not to be the case, however. Examples are given, in [69], Birnbaum (1977), and Berger (1983), of apparently irreconcilable differences that exist between the answers provided by the two approaches. For instance, non-frequentist conditional approaches to testing two simple hypotheses would state that the evidence to be reported is the likelihood ratio of the two hypotheses, for the observed data. Examples can be constructed (see Berger (1983)) where these lieklihood ratios seem unlikely to bear any similarity to any conditional frequentist reports. Furthermore, there is the likelihood principle (a very general version and review of which is given in Berger and Wolpert (1984)), which is based on believable axioms and contraindicates any frequentist averaging.

The problem stated at the beginning of Section 1 has, thus, not gone away. In fact, it now seems clear that one really has to make a choice: either abandon frequentist justification as the primary necessity (although frequentist measures could still have a number of extremely important roles to play in statistics), or be willing to accept the fact that, from time to time, any valid frequentist answer will violate "conditional common sense." Kiefer clearly states in [69] that he considers frequentist justification to be paramount, and that he is willing to live with the consequences. Indeed in [69] he presents a number of interesting philosophical comments on these issues. For a discussion of the "other side" see Berger (1983) and Berger and Wolpert (1984) (which reference earlier works on the issue).

In any case, the conditional frequentist approaches of Kiefer deserve widespread attention. Frequentists have an enormous amount to gain by implementing them, both in attaining closer correspondence to "conditional reality" and, in some cases, because of the great simplifications that result. Even Bayesians can take an interest in this work, since it is fairly well established that Bayes procedure that have good frequency (or perhaps, now, good conditional frequency) properties, tend to be robust with respect to prior specification. This work has opened new vistas in statistics. It would be a shame if statisticians refuse to enjoy the view.

References

Berger, J. (1983). The frequentist viewpoint and conditioning. *Proceedings of the Berkeley Symposium in Honor of J. Kiefer and J. Neyman* (L. LeCam and R. Olshen, eds.), Vol. 1, 15–44. Wadsworth Publishing Co.: Belmont, CA.

Berger, J. and Wolpert, R. (1984). *The Likelihood Principle: A Review and Generalizations.* IMS Lecture Notes Monograph Series #6, Hayward, CA: Institute of Mathematical Statistics.

Birnbaum, A. (1977). The Neyman-Pearson theory as decision theory and as inference theory: with a criticism of the Lindley-Savage argument for Bayesian theory. *Synthese*, 36, 19–49.

Brown, L. D. (1978). A contribution to Kiefer's theory of conditional confidence procedures. *Ann. Statist.*, 6, 59–71.